It is a warm June day. "We are all set to go!" says Dad. "Will you twins please carry your packs?"

"You bet!" yell Bart and Barb.

"It is quite far to Gram's farm," Dad says.

"We can play fun games!" says Bart.

"We have art stuff to do!" says Barb.

Mom asks, "Will you two please entertain Skylar? Then she won't start crying!"

Bart and Barb look at each other. Crying is NOT part of the fun!

First, they play peek-a-boo.
Bart and Barb put their hands
over their eyes.

Next, they make farm animal sounds. Barb says, "Baa! Baa! Baa!" Skylar rewards her with parts of her apple.

Then, the twins get out the art stuff. Bart makes a monkey puppet. He hangs it on his arm.

At last, Gram's red barn is in sight!
"BARN!" yells Skylar. And her eyes start to close.

"Look at Skylar," says Gram. "She is such a smart girl! I bet she slept for the whole trip."